Kids & TECHNOLOGY
MISSION 21

SPACE COLONIZATION

SHARON A. BRUSIC

Developed by Virginia Polytechnic Institute and State University. Funded by a grant from NASA.

DELMAR PUBLISHERS INC.®

For my mother, Darlene J. Rutkowski

With special thanks to Jodi Bergeman for her valuable research assistance and William E. Dugger, Jr. and James E. LaPorte for their roles in making this Mission 21 series possible.

Photography Credits
pp. 3, 4, 5, 6, 7, 8, 10, 11, 12, 14, 15, 16, 18, 19, 23, 24, 25, 26, 28, 29, 30, 31: Courtesy of National Aeronautics and Space Administration (NASA); p. 9: Photo by Sharon A. Brusic; p. 13: Courtesy VELCRO USA; p. 17: Courtesy Huth/*Construction Technology*, 2nd edition, copyright © 1989 by Delmar Publishers Inc.; p. 20: Courtesy Hoechst-Celanese Corp.; p. 22: Courtesy MiniMed Technologies, Inc.

Cover photo: *Courtesy of NASA*

Delmar Staff
Associate Editor: Christine E. Worden
Freelance Developmental
    Editor: Cynthia Haller
Project Editor: Laura Gulotty

Production Supervisor: Karen Seebald
Design Supervisor: Susan C. Mathews
Photo Research
    and Acquisitions: Michael Nelson

For information, address Delmar Publishers Inc.
2 Computer Drive West, Box 15-015
Albany, New York 12212

Printed in Canada
Published simultaneously in Canada
By Nelson Canada
A division of The Thomson Corporation

10 9 8 7 6 5 4 3 2 1

ISBN: 0-8273-4102-4

CONTENTS

THE FINAL FRONTIER

Imagine living in space for months at a time. Your home would be a **space station**, a giant structure orbiting Earth. This picture is one artist's idea of a space station.

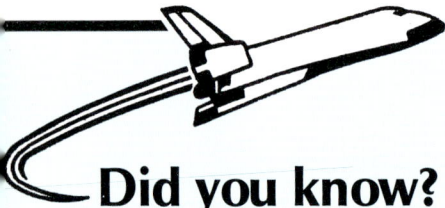

Did you know?

America's first space station was called Skylab. It went into space in 1973. Skylab was the size of a 3-bedroom house. It orbited Earth about 270 miles above the planet. Three different crews lived and worked in Skylab at different times. They spent a total of 169 days there. They studied Earth, space, and how living in space affects people.

People have explored every part of Earth. They have gone everywhere, from deep under the sea to the cold polar regions to the upper reaches of the atmosphere. Now we are moving into space. Some people call space the "Final Frontier." It is the last place left to explore. It is also our greatest challenge.

Seek, Find, Create

In 1984, President Ronald Reagan directed the **National Aeronautics and Space Administration (NASA)** to start working on a **permanent space station.** This will be a place where people will live and work full time in space.

From a space station, scientists will be able to study both Earth and space. They will look down on Earth and study its weather and oceans. They will learn more about changes in climate that affect our whole planet.

Learning about space will be easier from a space station, too. Earth's atmosphere makes it hard to study the sun, the stars, and the planets. From a space station, scientists will get a better look.

The goal of the space station is to *seek, find,* and *create.* People in space will seek new knowledge and find new information. They will create new tools and new materials. People everywhere will benefit from what is learned and created.

Moving into Space

The space station is only a beginning. Someday, there will be cities in space. People will live in space. **Space colonization** is the term for living in space. They will be born, grow up, and work in space. Some kinds of jobs can most easily be done in space. Must a satellite be repaired? Must it be moved into a new orbit around Earth? Those are jobs for a space worker.

Space industrialization means working and producing in space. Space factories will make products that can only be made in space. Most important, space stations will provide a jumping-off point for space exploration. It will be easier to go to the moon and the planets from a space station. Spaceships will be built in space. They will leave from space stations and come back to them at journey's end.

Living and working in space will mean solving many new problems. We will learn how space affects the human body. We will learn to grow food in space. We will learn the best ways to reuse scarce resources such as water and air. We will develop ways of powering machines in space. We will find out what materials can be used to build space structures. We will discover the best ways to put them together.

In solving these problems, people will gain new knowledge. This new knowledge will be used not only by people in space, but by people on Earth.

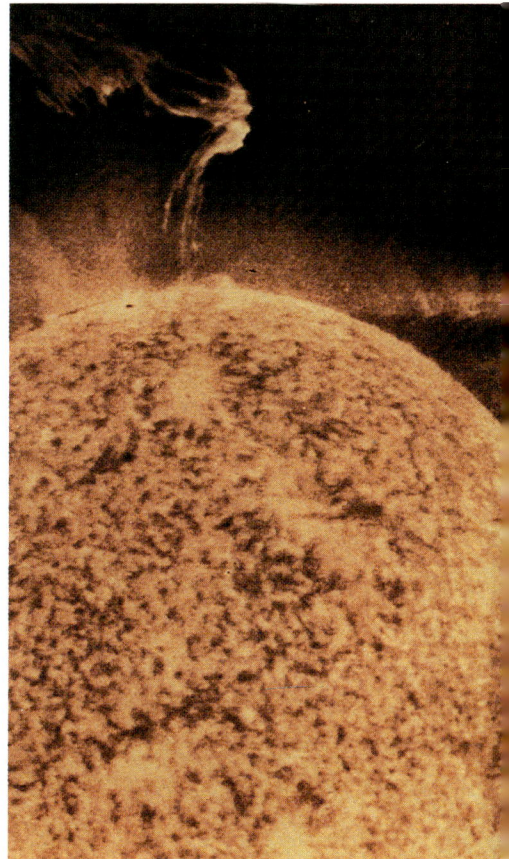

America's first space station was Skylab. Skylab's scientists learned much about the largest body in our solar system. What is it? (Answer on page 32.)

FREEDOM STATION

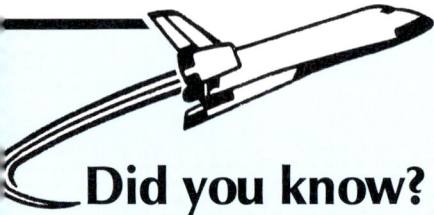

Space Station Freedom is the name given to the first permanent space station. *Freedom* may be ready by the year 2000. People will travel to the space station by **space shuttle**. They will live and work there full time, all year round.

A Place in Space

Freedom will not be built on Earth and sent into space. Instead, small pieces of it will be built on Earth. Some pieces will be sent into space by rocket. Other parts will go aboard space shuttles. **Astronauts** will go into space on the shuttles, too. They will put the parts together to make the space station. Astronauts will wear jet-powered backpacks called **manned manuevering units** or **MMUs**. MMUs will help the astronauts move about in space.

First, a framework of aluminum will be prepared. Then **modules**—air-tight structures in which astronauts will live and work—will be attached. Each module will fit into the payload bay of a space shuttle. In this way, modules can be brought to the station ready to use. There will be several kinds of modules.

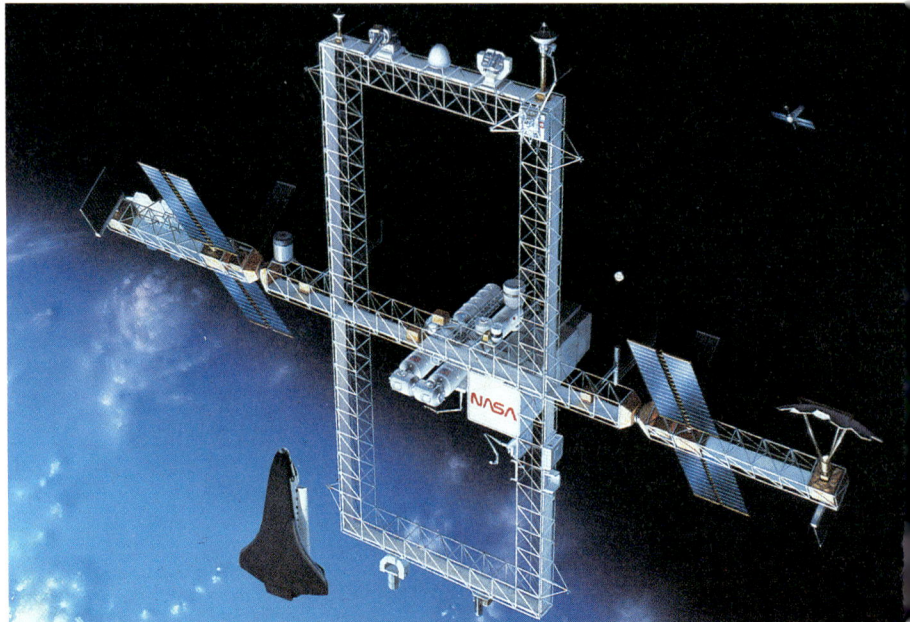

The space station will have many parts that serve different purposes.

1. *Living quarters.* This module will have room for up to eight astronauts to live and sleep.
2. *Science lab.* Scientific studies will be carried out in this module.
3. *Materials processing lab.* Here, ideas for space products will be tested and studied.
4. *Storage compartment.* The supplies and materials needed by the space station will be stored in this module.
5. *Service center.* This module will be used for repairing **satellites**.

Another important part of *Freedom* is its power station. The power station will be made up of two huge panels of **solar cells**. Each will be the size of a football field. These solar cells will use sunlight to make the electricity needed to power the space station.

When the space station is complete, astronauts will go in groups to spend three months living and working on *Freedom*. Space shuttles will bring supplies and materials as they are needed.

A Stepping Stone

The space station will be built by many nations. Japan, Canada, and the European Space Agency (ESA) are all part of the project. (ESA is like NASA, except that ESA represents 13 nations. NASA represents only the United States.)

This joint participation makes *Freedom* a very special project. The cost will be shared. The problems will be solved by people from many nations. Ideas will be shared. *Freedom* will help to bring these nations closer together.

The space station could make it possible for people to reach this planet by the year 2020. What planet is it? (Answer on page 32.)

ZERO GRAVITY

Astronaut Joe Allen is playing with his orange juice. In space, liquids do not stay in open containers, but float as balls or *globules*.

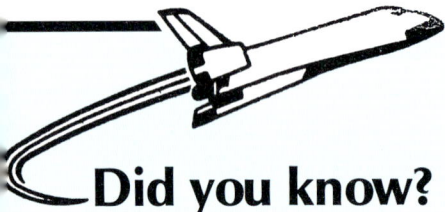

Did you know?

If you ever have to write a letter underwater, you can use a pen invented for the space program. On Earth pens work because gravity makes the ink flow. Astronauts needed a pen that would write in the zero-gravity conditions of space. Paul Fisher of the Fisher Pen Company invented the **space pen**. It uses a rubberlike ink. It can write at any angle. It works great in space or underwater! Divers and astronauts like the space pen.

Gravity is the unseen force that pulls things toward the center of Earth. If you push this book off your desk, it will fall to the floor. A thrown ball always falls to the ground.

Aboard the space shuttle, astronauts must deal with **zero-gravity** conditions. The gravity of Earth pulls on the shuttle and everything inside it. Because the shuttle is *orbiting,* though, there seems to be no gravity. People and things are **weightless**. Aboard the space shuttle, there is no real "down." If you want to stay on the floor, you have to attach yourself to it somehow. Water will not go down a drain. Taking a shower or using a toilet requires special equipment. Tools do not stay where they are put.

Weightless Wonders

Weightlessness is troublesome—and useful. It's a problem for people who live and work in space.

It's not hard to get used to being weightless. But weightlessness is hard on a person's body. People lose muscle and bone mass in a short time. Astronauts have to follow an exercise program in space to keep their bones and muscles strong.

Every task of living and working becomes a challenge in weightless conditions. Just getting where you want to go is a job. Objects can't be put down. They must be stuck somewhere if you want to find them again.

On the other hand, weightlessness can be useful.

Substances in space act differently than they do on Earth. For example, consider **crystals**. If you put table salt in a glass of water, it dissolves. After a few days, the water will evaporate. Crystals of salt remain in the shape of cubes. Salt crystals *grow* as the water evaporates.

On Earth, gravity affects this process. Crystals sink to the bottom of the glass. They bump each other. They are squashed against the glass. They cannot grow into perfect cubes. On the space shuttle, crystals float in space as they grow. Their crystal shape is perfect. Crystals can grow larger in zero-gravity conditions. On Skylab, crystals of a protein were grown. They were 1,000 times as big as any grown on Earth. These jumbo crystals were taken back to Earth. They were used to study the protein.

Zero gravity can be helpful in other ways. Think about how costly it is to get away from Earth. Spaceships must use tons of fuel to overcome the pull of gravity. If there were no gravity to overcome, getting a spaceship moving would require a lot less fuel. If a spaceship could be built on a space station orbiting Earth, it could be launched at much lower cost.

Growing crystals for research and low-cost space travel are just two ways to use zero gravity. There are many more. Who knows what the future will bring?

Astronauts must get used to living and working in weightless conditions. Where on Earth can they do this? See the photo above for a clue. (Answer on page 32.)

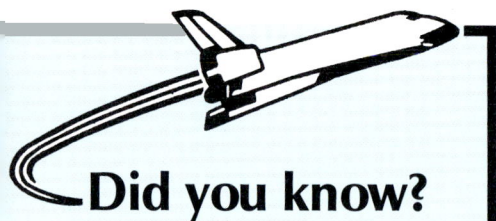
The space shuttle is part of the Space Transportation System.

3...2...1... LIFTOFF!

Have you ever watched a space shuttle launch on TV? It's exciting when the rocket engines fire and the shuttle begins to move. Imagine how the astronauts feel! Perhaps you wish you could go with them as you watch the shuttle soar above the clouds.

Once, space travel was something to dream about. Writers told stories about it. Today, it's reality. People are really traveling into space.

The Space Routine

The space shuttle is the first reusable spacecraft. Past space vehicles, like those in the Apollo and Gemini programs, went into space only once. Each trip required a new vehicle.

The space shuttle changed that. Its design lets it take off like a *rocket*, orbit Earth like a *spacecraft*, and land like a *glider*. Also, it can transport things into space like a cargo plane. It can carry satellites and equipment for experiments. Best of all, the shuttle can do these jobs over and over again.

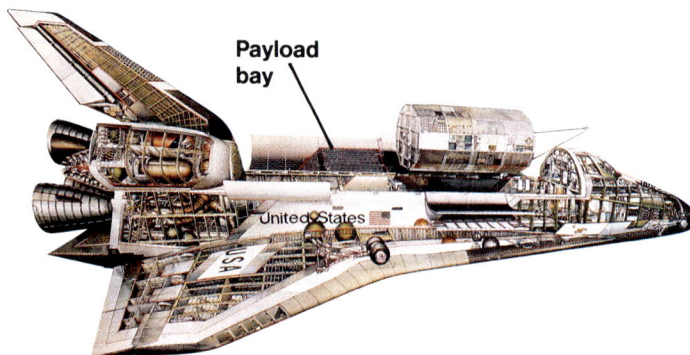

More Than a Machine

The space shuttle is more than a machine. It is a **system**. In a system, many parts work together to do a job. The shuttle system is called the **Space Transportation System (STS)**. Although there are hundreds of parts in the STS, four parts are most important:

Space shuttles weighing up to 240,000 pounds land on the surface pictured above. What is it and where is it? (Answers on page 32.)

$$\begin{array}{rl} & 1 \ \text{orbiter} \\ + & 1 \ \text{external fuel tank} \\ + & 2 \ \text{solid rocket boosters} \\ \hline = & 4 \ \text{important parts of the space shuttle} \end{array}$$

Look at the picture of the space shuttle on page 10. You will see that the orbiter is the major part of the STS. It contains the **crew compartment**. This is where the astronauts stay during most of a shuttle flight. It also contains the **payload bay**. This is the shuttle's storage area. All kinds of cargo are kept here.

The payload bay has room for several satellites and a lot of scientific equipment. When a trip is planned, care is taken to fill the payload bay but there is often space left over.

In the little leftover corners, small containers called Getaway Specials are placed. For a few thousand dollars, a person or company can buy shuttle space — a Getaway Special. The Getaway Special containers come in two sizes. One holds experiments up to 5 cubic feet and 200 pounds. The other is half that size. Whatever goes inside must run by itself. An astronaut turns on each Getaway Special once the shuttle is in orbit.

The orbiter has three **main engines**. They are fueled by the huge **external tank**. This part of the shuttle is not reused. After the fuel is gone, the shuttle drops the external tank into the ocean.

Solid rocket boosters (SRBs) help push the space shuttle into the sky about 28 miles from Earth. Then they separate from the orbiter and fall into the ocean. Ships find them and they are pulled from the water. They are reused on future flights.

"HOME SPACE HOME"

Would you like to *sit on the ceiling, sleep on the wall, and eat standing upside down?* You could do all those things and more on the space shuttle.

On the shuttle, even the simplest tasks of living have to be carefully planned. Weightlessness makes life very different. On Earth, gravity keeps pulling you down. In orbit around Earth, you float. It sounds like fun, yet a number of inventions have been necessary in order for people to feel at home in space.

Cooking and eating are both problems in space. Food must be stored in a way that will take up as little room as possible. It must be packaged in a way that will keep it from getting loose and floating off. Loose materials of any kind are likely to be sucked into the air filters and clog them.

Some of the food eaten on the shuttle is stored in **rehydratable packages**. These are plastic packages

In space, it's hard to tell whether you are standing on your head or standing on your feet.

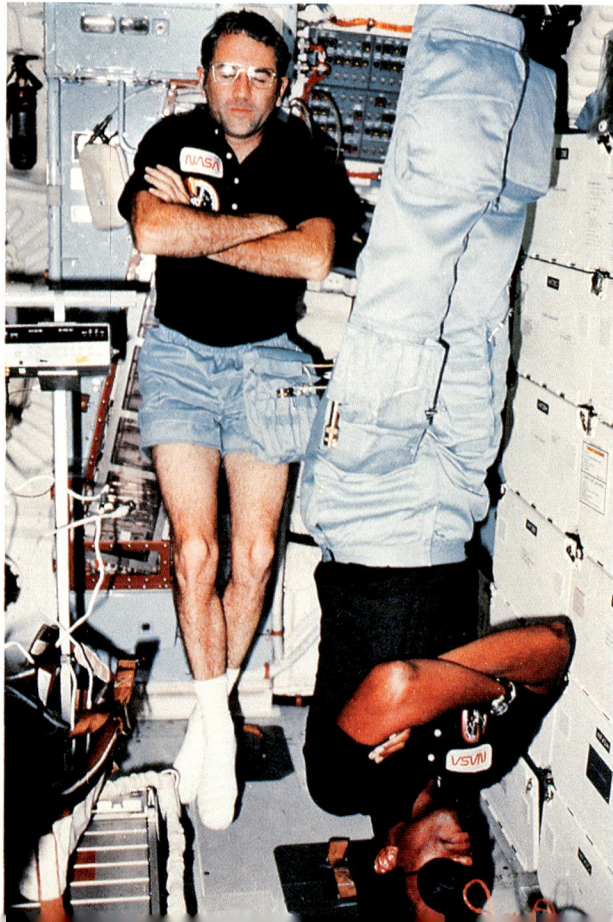

Did you know?

Everything is different in space, even seasoning food. In weightless conditions, you cannot shake salt and pepper onto your food. The seasonings would float through the air. Instead, you have to squeeze liquid salt and pepper into your food pouch. Space salt is dissolved in water. Space pepper is mixed in oil. Even small tasks become harder in space.

of dried food to which water can be added. Water is added through a needle to prevent leaking. Other packages contain solid foods that can be warmed in an oven.

Meals are placed on lap trays and eaten with knives, forks, and spoons. Liquid foods must be sucked through a straw. The straw has a clamp. When it is not being used, the straw can be closed off so no liquid will escape. Astronauts are careful not to drop food or bump each other while eating. If they do, they must get out a vacuum cleaner and clean up their spills immediately.

Toilets are also different in space. Astronauts use a seatbelt and hold onto handles so they don't drift away from the toilet. The toilet itself uses suction — pulling streams of air — to get rid of waste. Astronauts wash their hands using a special enclosed washer. They stick their hands in and the washer sprays water. Used water is sucked into a storage tank.

Astronauts must sleep and exercise. In space, both these tasks are a little different. A common exercise is running in place. To keep from moving around, astronauts wear straps or belts as harnesses. Astronauts sleep in sleeping bags. They can drift free or attach their sleeping bags to the walls.

Other Problems, Other Solutions

Astronauts must take everything they need with them into space. Space is airless, so a supply of air must go on the shuttle with the astronauts. It must be filtered to keep it clean. It must be conditioned to keep the temperature just right.

The astronauts must take medicines, in case of illness. All their food and water must go with them. Many tasks that are simple on Earth are problems in space.

Technology helps to solve these problems. Technology is using knowledge and materials to provide for people's needs. In space, even the simplest needs can take quite a bit of thought and planning.

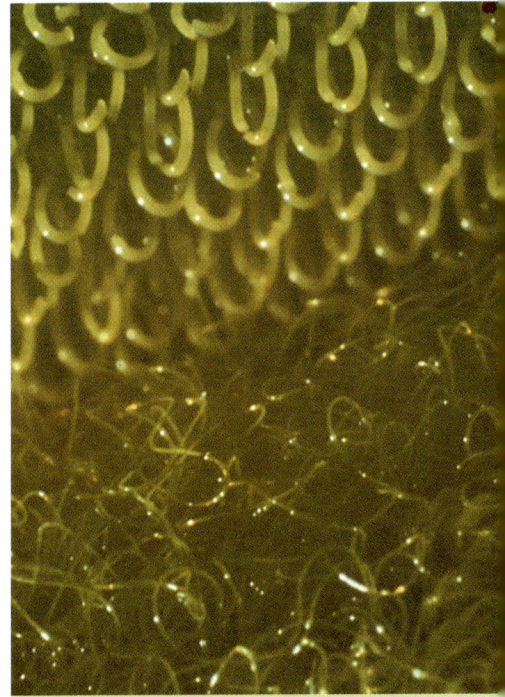

What fastener helps astronauts keep objects from drifting away in the space shuttle? The photo above may give you a clue. (Answer on page 32.)

13

MADE IN SPACE

This machine is used for containerless processing of glass.

![shuttle illustration]

Did you know?

Because of a student, 3,300 extra passengers—all of them bees—took a shuttle flight in 1984. The student designed an experiment to find out if bees could build a honeycomb in zero-gravity conditions. The student discovered that the bees were confused at first. Later, though, they got busy and built a honeycomb.

You see tags that say "Made in Japan" or "Made in the U.S.A." You've never seen one that says "Made in Space," but perhaps someday you will. Plans are already underway to build **space factories** that will orbit Earth. Medicines and computer materials will probably be among the first space-made products. When a tag says "Made in Space," you'll know you have something that was made in a space factory.

Space Factories

Picture this. An astronaut traveling in space squeezed a little water from a plastic bottle. The water became a perfectly round ball floating in air. That ball of water is one example of something that can be made in space that cannot be made on Earth. Zero-gravity conditions make it possible.

Besides perfect spheres, what else can be made in space? **Alloys** can. Alloys are mixtures of different metals. Two or more metals are melted and mixed to make a new metal. Brass, for example, is a mixture of copper, zinc, and other metals. An alloy is different from the metals mixed to make it. It can have new and different uses.

On Earth, some metals don't form alloys. They won't mix when melted. The lighter metal floats on top of the heavier one, like oil on water. This happens because of gravity. In zero-gravity conditions, however, mixing is easy. All kinds of new alloys can be made. These will be metals unlike any made on Earth. It may be possible to create a metal to solve a certain problem. New metals may find many new uses.

No Pots or Pans Needed!

Another space factory job will be **containerless processing**. To *process* something means to change it in some way. Melting, cutting, or mixing are ways we process materials.

On Earth, a container is needed for many kinds of processing. To make glass, sand and other materials are heated *in a container*. In space, this can be done without a container. The glass does not have to touch any surface. That means that tiny bits of the container's surface don't get mixed into the glass.

Glass made in this way is very pure. It is much purer than it would be if it were made on Earth. A space factory could make large amounts of such **ultrapure** glass. The glass could be used in special instruments. Other ultrapure materials could be made in this way, too.

New alloys and pure glass are just a few of the many things a space factory could make. As time goes on, people will have more ideas about products that can be made in space.

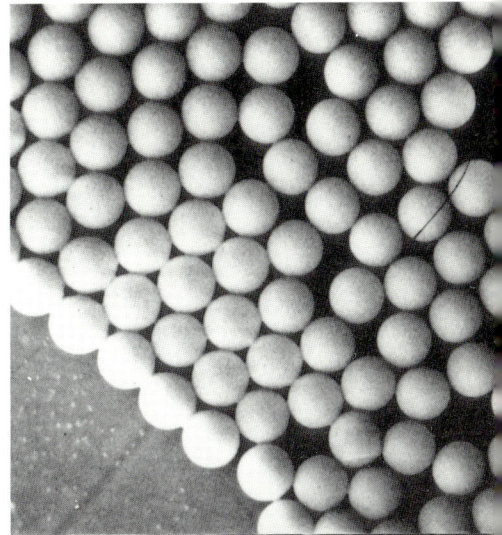

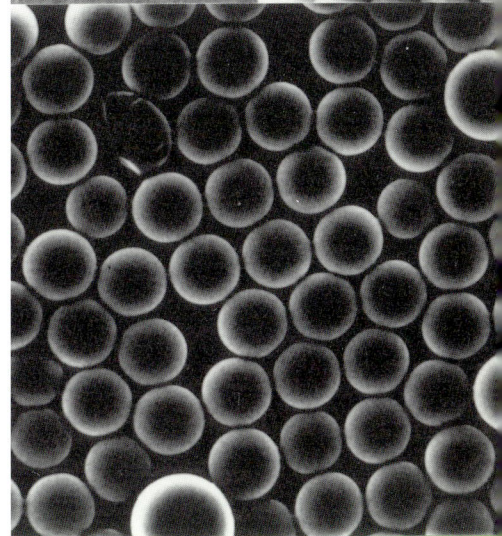

The spheres in these pictures help people to measure very tiny objects. They are used by companies that make cosmetics and paint. Which picture (top or bottom) shows the spheres that were made in space? (Answer on page 32.)

15

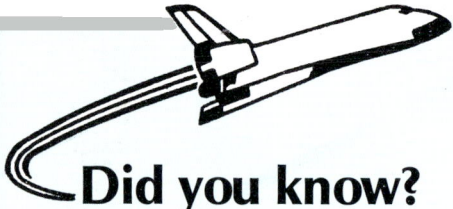

You have probably played with *Tinkertoys*® or *Legos*®. They are building sets that have small parts that fit neatly together. In planning the space station, designers have used this idea. The space station will be made up of parts that can be easily fastened together.

Building a space station will be much harder than building anything on Earth. Astronauts wearing space suits will have to do most of the job. Each step of their work must be carefully planned ahead of time.

From Struts to Modules

Space station construction will start with a framework on which other parts will be fastened. The framework will be made up of **booms** up to hundreds of feet long. The booms, in turn, will get their shape and strength from **trusses**. The trusses

Astronauts will build the space station.

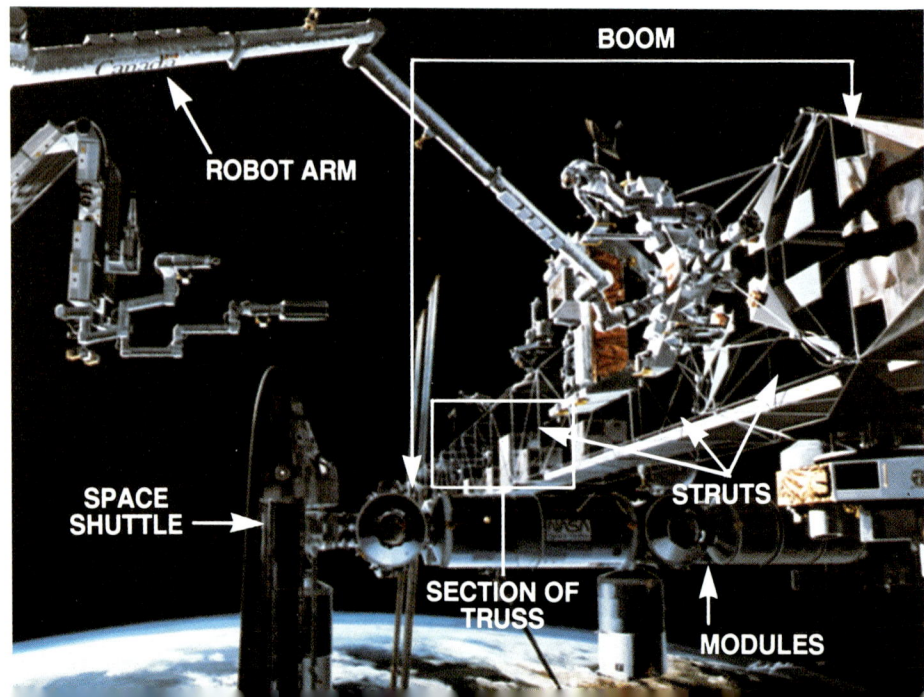

ROBOT ARM
BOOM
SPACE SHUTTLE
STRUTS
SECTION OF TRUSS
MODULES

16

will be made of long, lightweight pieces of metal called **struts**.

Space living and working areas are called modules. The modules will be built on Earth. They will be moved into space in the shuttle's payload bay. Astronauts will use **robot** arms to help fasten the modules to the space station's booms. They will put the booms and modules together as if they were building a giant *Tinkertoy* model.

Building in Space

Think about building a *Lego* model underwater. That will give you an idea of some of the problems the astronauts face in building the space station. They will have to carry their own supply of air with them into space. They will have to wear large, bulky suits and helmets for protection. The astronauts will move carefully, planning each motion ahead of time. They must have all the tools they need with them. Going back to the shuttle for something would be a lot of trouble.

The astronauts will have to go through months of training. They will have to learn to work in zero-gravity conditions. They will also have to learn building skills. They must be the kind of people who can think quickly and solve their own problems. Help will be far away.

Some devices may be used to make the job a little easier. Astronauts won't have to pull themselves along booms, hand over hand. Instead, they may ride along the booms on a flatcar.

Another way the astronauts will move around is with a **manned manuevering unit (MMU)**. This is a jet-powered backpack. It works by letting off a tiny amount of gas under pressure. Think about how a blown-up balloon flies around the room when you release the air. That will give you an idea of how an MMU works.

You can see that building the space station will be a big job. With careful planning, it can—and will—be done.

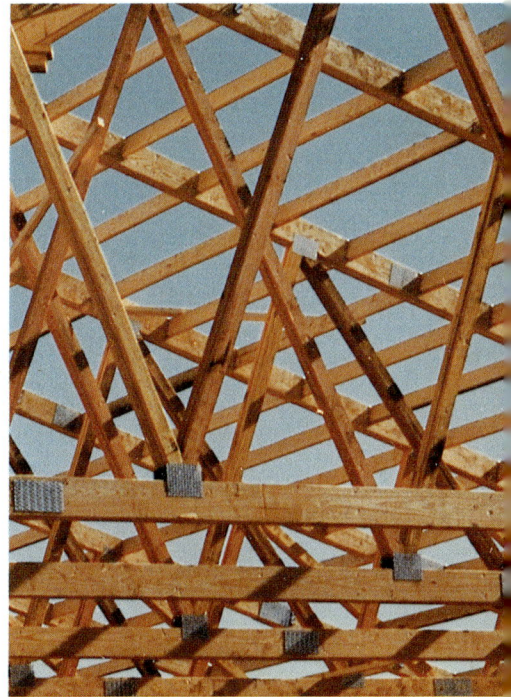

Many kinds of buildings use **trusses** like those that will be used to build the space station. The trusses in the photo above are part of a building on Earth. What part is it? (Answer on page 32.)

Did you know?

Robots get their name from a word in the Czechoslovakian language. The word is **robota** and it means slave-like work, or forced work. Robots work on Earth and in space. They can do jobs that would be dangerous or too difficult for people. They are helping us explore the solar system. They will play an important part in building and running a space station.

The robots you see in movies aren't much like real everyday robots. Movie robots are made to look like people, with a body, head, and arms. They communicate by talking or beeping. Most real, working robots represent just one part of a person—the arm. They are controlled by computer.

If you have ever played a computer game or video game, you know about computer control. You push the buttons, and that controls the game computer. The computer, in turn, controls what happens on the TV screen.

In a robot, much the same thing happens. A person programs the computer. The computer controls every move the robot makes.

Super Workers

Robots are super workers. They can work all day, every day, with no breaks. They don't get sick.

The diagrams at left show all the ways a robot arm can move. Which can you see in the picture of a robot arm in space?

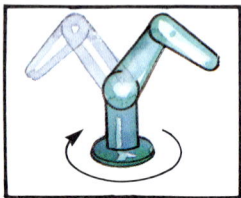
Waist

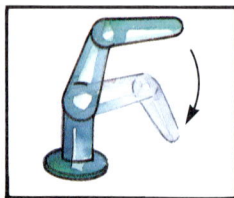
Shoulder

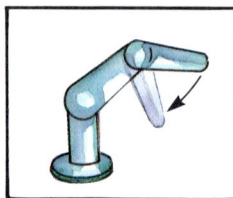
Elbow

Roll

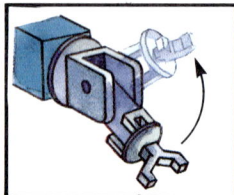
Pitch

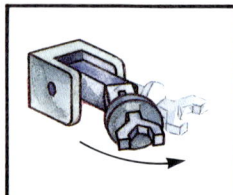
Yaw

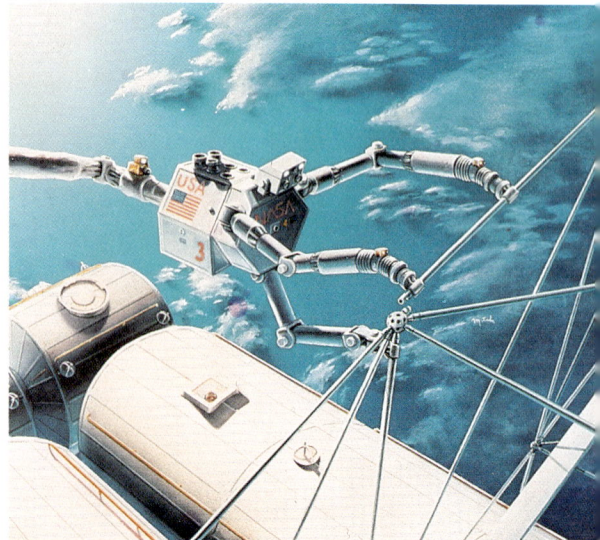

Robots do the same job over and over again, perfectly, without getting bored. They can work in intense heat or surrounded by paint fumes. They can work at the bottom of the ocean or in airless space. A robot can pick up 1000 pounds. It can place an object into a red-hot furnace.

Robots are also stupid workers. They can do only what they are programmed to do. If a job changes, a robot must be reprogrammed. If something goes wrong, a robot cannot fix it. The robot will just keep on doing the job it was programmed to do until someone turns it off.

Arms in Space

No person has ever set foot on a planet other than Earth. Robots, however, have already traveled to Mars. In 1976, the two Viking spacecrafts landed on Mars. Robot arms aboard each ship scooped up soil and rock. They helped carry out tests on these materials.

On the space shuttle, astronauts use a robot arm called the **remote manipulator system (RMS)**. This robot arm is used to pick up satellites and bring them into the shuttle. It has cameras and a special grip. The RMS is useful in handling large objects like satellites.

There is no limit to the jobs robots will do in space. They will be used to build the space station. A robot arm can move bulky parts and hold them in place. Just as on Earth, robots will probably be used to work in space factories. They may be used to move and process materials. They may also be used to package materials and prepare them for the return trip to Earth. In the future, robots may explore places in the solar system that are too dangerous for people.

Robots provide a new way to do jobs that are difficult (or even impossible) for people. Whether they are at work on Earth or in space, robots show how technology can be used to provide for human needs.

The Voyager II space probe sent this image back to Earth. What is pictured here? (Answer on page 32.)

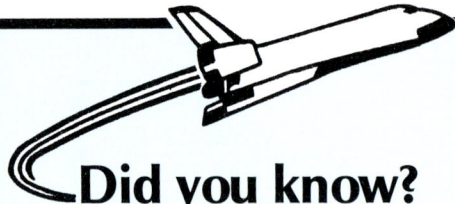

Did you know?

A system invented for the space program protects the Declaration of Independence in Washington, D.C. The system was invented to make pictures of bodies in space like planets and moons. This system is used on Earth to make computer images of the Declaration of Independence. The system checks the images to make sure everything is all right. It can find problems before people are able to see them. It can tell if paper is splitting or ink is fading. People can fix the problem right away before it gets worse.

What's in it for me? That is a question that many people have asked about space technology. A lot of money and hard work have gone into the space program. What are the benefits for people who will never travel on the shuttle?

The answer is: space **spinoffs**. Space spinoffs are new products, processes, and ideas that have come from the space program. Space spinoffs are an example of **technology transfer**. Technology used or invented for one job can be used to do jobs in other areas.

New Materials

Many new materials have come from the space program. In the 1970s a special fabric was invented to protect instruments from the heat and cold of space. It is plastic coated with a thin layer of metal. Today this material— *TEXOLITE*®— is used in sleeping bags, parkas, and boots. It keeps peo-

The firefighter's jacket is made of *PBI*® fibers. The fibers protect him from the fire's extreme heat.

ple warm. *Texolite* is also used to line window curtains and blinds. This helps to save fuel by keeping homes warmer in winter.

Another material was invented for the flight suits of astronauts. NASA wanted a material that would provide protection in case of a fire. A fiber called *PBI*® was invented to to this job. Today *PBI* is used in suits for firefighters. It will withstand intense heat and fire. If you've ever traveled by plane, you may have been sitting on *PBI*. Airlines cover their cushions with this fabric. If a fire breaks out, the fabric keeps the foam in the cushion from burning. If the foam were to burn, it would release harmful gases into the air.

New Products

One useful machine was developed because NASA wanted to learn about the buildup of ice on spacecraft. The machine that was developed measured very tiny amounts of particles. Using that machine as a model, people built the *Particulate Mass Monitor*®. It can be used to measure small amounts of particles in the air. It is useful in checking the air for smoke. It can also be used to find out if engines that burn fuel are working well.

Aboard the space shuttle, water must be reused. A special water filter cleans water so it can be used again. That filter provided the idea for a whole new company. An Arizona man didn't like the taste of the water where he was living. Using NASA's filter idea, he built a water filter for his own home. Friends and neighbors wanted one, so he started a company to manufacture the filters in large numbers.

There are many other space spinoff products. You may have seen battery-powered screwdrivers, vacuum cleaners, and drills. The idea for battery-powered tools came from the Apollo program. (It was the program that landed people on the Moon.)

One goal of the space program is to share new ideas and information.

What space spinoff, invented for astronauts, is used by hospital patients to clean their teeth? There's a clue in the picture. (Answer on page 32.)

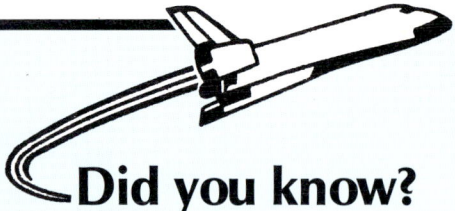
This boy is wearing the *MiniMed® 504*. This device (about the size of a credit card) helps diabetics·lead a more normal active life.

MEDICAL MARVELS

You depend on medical technology in many ways. You get *shots* to keep you from getting diseases. If you break a leg, *X-rays* will be taken of your bones. Dentists use *special tools* for cleaning and filling your teeth. Other doctors use *eyeglasses* or *contact lenses* to help you see better.

Medical technology is getting better all the time. The space program deserves credit for some of this progress.

Tiny Wonders

People with diabetes have bodies that do not produce enough **insulin**. Insulin keeps the amount of sugar in the blood at a normal level. If this level is not kept normal, a person becomes weak and can die. In the United States, more than a million people have diabetes. They must take daily shots of insulin.

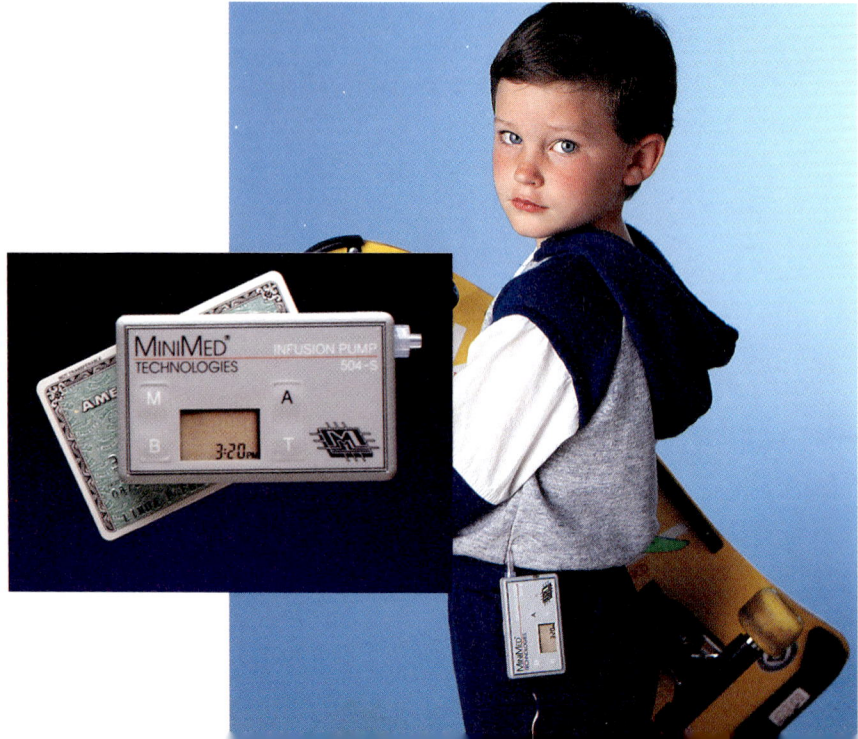

People with diabetes, called *diabetics*, must be careful in the way they live. Different kinds and amounts of food and their level of exercise affects their need for insulin. They must plan carefully to keep everything in balance.

A new device, *MiniMed® 504*, lets diabetics lead a more normal life. The device is about the size of a credit card. Battery-powered, it clips to clothing.

Part of the device is a tiny pump that pumps insulin into the body a little at a time. Part of it is a computer. The computer is programmed to provide insulin to the body according to the person's needs. This means that the amount of insulin a person gets is just right. Many diabetics have eye and kidney problems because they cannot balance insulin needs. The *MiniMed 504* will help to prevent these problems.

The technology used in the *MiniMed 504* was first used on small satellites. Some of these satellites were the size of a beach ball. A very tiny control system was needed. Because the *MiniMed 504* is so small, diabetics find it easy to use.

NASA research led to the use of a plant to treat wastewater. It is used today in sewage treatment at Walt Disney World in Florida. What is the plant? (Answer on page 32.)

Many Marvels

Over the years, many medical marvels have come from space research. Here are just a few:

- A device to help deaf people speak more clearly. It shows a pattern of normal speech next to a pattern of the deaf person's speech. The deaf person then matches his or her speech with the normal pattern.
- A device that can test vision in babies and very young children.
- Glue that works underwater. It can be used in filling teeth.
- A wheelchair that talks. A computer and *voice synthesizer* are used to help a person who is unable to speak.

People use space technology to come up with great ideas. Some of them really are medical marvels.

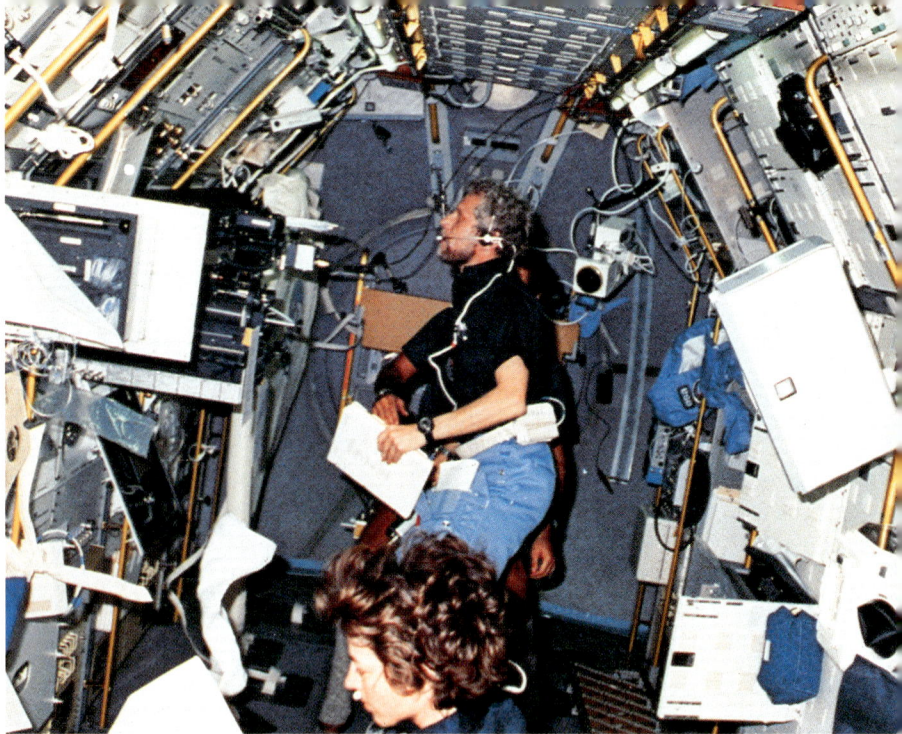

Space is limited in shuttle living quarters. Astronauts must get used to tight spaces like this.

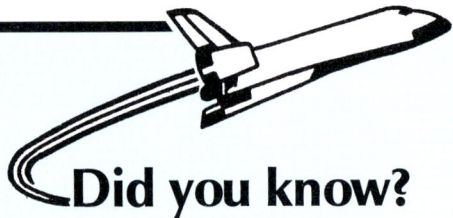

Did you know?

If you need to have fresh fruit, don't become an astronaut. Shuttles cannot carry heavy refrigerators to store a lot of fresh food. Fresh fruit spoils rapidly. Astronauts must eat applesauce, not apples. They have a grape drink, not grapes. Much of the fruit they enjoy is canned or dried, not fresh.

How much would you be willing to pay for a trip into space? Even if you're a millionaire, you couldn't afford it! Space travel is expensive. When you add up all the costs of putting people into space, the bill is huge.

Some people think space travel is important. They believe that they get many benefits from the space program. They believe that we can learn much from space travel. Others think money is being wasted on space programs. They would prefer to spend the nation's money in other ways.

Paying the Price

Space travel costs billions of dollars. In the United States, people pay for the space program through taxes. A tax is money paid to the government by citizens. It is based on how much a person earns. The government spends the money it collects on programs and services.

Some tax money is spent on education. Some is spent on *defense*, to protect the country from attack by other countries. Some money goes for health care and helping needy people. Some is spent to build roads and bridges. A small amount goes to the space program.

People have many needs in the United States. The government cannot collect enough money to take care of all these needs. It must decide where tax money will be spent. Many choices must be made. Spending decisions are difficult decisions.

More Than Money

Sometimes, space exploration costs more than money. Sometimes, astronauts give their lives. In 1986, seven astronauts died in a space shuttle accident. Many jobs are risky. At this time, being an astronaut is one of them. In the future, this job may not be so risky.

Space living may be harmful to a person's health. One problem is living in a small space, close to other people. Astronauts have no privacy. They must eat, sleep, and work near others. Doctors already know that *weightlessnes*s affects bones and muscles. They have questions about long-term space travel. Nobody knows how long people can safely live in space.

There's no doubt that the space program is costly. People ask whether the time and money being spent on space would be better spent on people on Earth. They wonder if the skills used in solving space problems could be used to solve problems that affect many Americans.

Do the risks and the costs outweigh the benefits of our space program? Think about it. Someday, you will pay taxes. Then, you will have to help make that decision.

It cost billions of dollars to travel to the Moon. Astronauts took pictures while they were there that were never possible before. This is one of them. What is it? (Answer on page 32.)

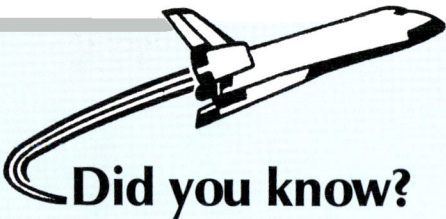

Did you know?

Where would you need an **aluminum fingernail**? In space, of course! Astronauts must wear spacesuits with heavy gloves when they go outside their spacecraft. It's hard to do jobs like picking up small parts when wearing these gloves. Aluminum finger-nails, fitted over the fingertips of spacesuit gloves, make these kinds of jobs easier. Sometimes small tools can make a big difference!

Astronauts need special clothing and tools to work outside their spacecraft. Here you can see an astronaut wearing a spacesuit and foot restraints.

At a hardware store you'll see many useful tools, materials, and machines. You'll see hammers, nails, cans of paint, and brushes. You'll see power tools, bins of screws, and stacks of lumber. A hardware store is the place to find what you need to make or fix things.

Now, imagine a hardware store in space. Think of it as a **spaceware store**. What will you find there? You'll see some things you recognize, but others will look strange to you. People need different materials and tools in space than they do on Earth.

Hold On!

Many tools used on Earth work fine in space. Space workers must do things a little differently, though. For example, when you use a screwdriver, gravity holds you down. You don't have to worry about pushing yourself away from the screw as you turn it. That would be a problem for a space worker using a screwdriver. In space, this problem is solved with **foot restraints**.

Foot restraints are important spaceware. Whenever an astronaut wants to stay in one place to do a job, he or she must use them.

Spaceware: Spacewear!

Spacesuits are another kind of spaceware. When painting a room, a person may wear a coverall and mask for protection. An astronaut often needs protection, too. Sometimes it's necessary for astronauts to go outside the spacecraft. People call this a space walk. To astronauts, it is **EVA**, or **Extra-Vehicular Activity**. An astronaut on EVA must wear a spacesuit. The spacesuit does many jobs:

1. It holds air under pressure.
2. It provides oxygen for breathing.
3. It keeps body temperature in a normal range.
4. It keeps air humidity (water content) at the right level. With too little, the astronaut's health may be damaged. Too much humidity causes water drops to coat the inside of the suit. This will cause the faceplate to fog and keep the astronaut from seeing clearly.
5. It protects against fast-moving bits of space dust, called *micrometeoroids*. The spacesuit protects the person from dangerous *radiation* from the sun and outer space.

Spacesuits are one of the most important items of spaceware. Without them, astronauts would have to spend all their time inside their spacecraft. If something went wrong, it would have to be fixed from the inside — or not at all.

How do special filters like the one shown in the picture help people living in space? (Answer on page 32.)

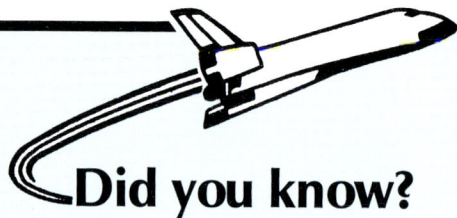

If you're a Scout, you have something in common with many astronauts. More than half of them took part in scouting as children and young adults. Kathryn Sullivan, the first American woman to walk in space, was a Girl Scout. Neil Armstrong, the first person to walk on the moon, was a Boy Scout.

Astronaut Dale Gardner has captured a satellite. He will take it to the shuttle for a return trip to Earth where it will be repaired.

HELP WANTED: SPACE WORKER

If you saw this help-wanted ad, would you apply for the job?

> **Space Worker:** Person to live and work in space for 90 days. Will test new space tools and help build addition to space station *Freedom*. Must pass physical exam. 12-month paid training period. Education and work experience required.

You don't have the qualifications to apply for this job now, but you could have them in time to help build *Freedom*. In fact, if you start preparing today, you could be one of the people who get the job. Tomorrow's space workers are children and young adults today.

Space Work

The space station will require thousands of workers. Although most of them will work on Earth, some will live and work in space. They will have experiences no one has ever had before.

After 20 or 30 years, *Freedom* will have had many space workers. They will have done all kinds

of jobs. Some will check and repair problems with the station's structure. Some will be in charge of unloading the shuttle and moving cargo to where it will be used or stored. The first crews will be small groups of no more than eight people. Later, there may be room for many more. These crews, whether big or small, will have to do all the jobs needed to operate the space station.

Teamwork

Many people in space and on Earth will work together to get the space station built and running. Here are a few areas in which workers will be needed.

Mission Control. Space shuttles will take people, materials, and tools to and from space. Teams of *controllers* make sure the shuttle trips are carried out as planned. They check all the systems of the craft. They keep track of crew members' fitness. A *flight director* is the person in charge at mission control.

Orbiter Processing. After each trip, the shuttle orbiter returns to Kennedy Space Center. There it is checked and repaired by technicians for its next trip. The two rocket boosters and the fuel tank are attached to the orbiter. The Space Transportation System is ready for another launch!

Engineering and Design. All the parts of the space station must be designed to work as well as possible. *Engineers* will design new parts and make sure they can be carried on the shuttle. *Technicians* will test them and put them together. *Technical writers* will write directions explaining how the parts work. *Training specialists* will show space workers how to use tools to build, repair, and operate the space station.

There will be many jobs for space workers in the future. Some of them will be in space. Many more of them will be on Earth. One of them could be yours!

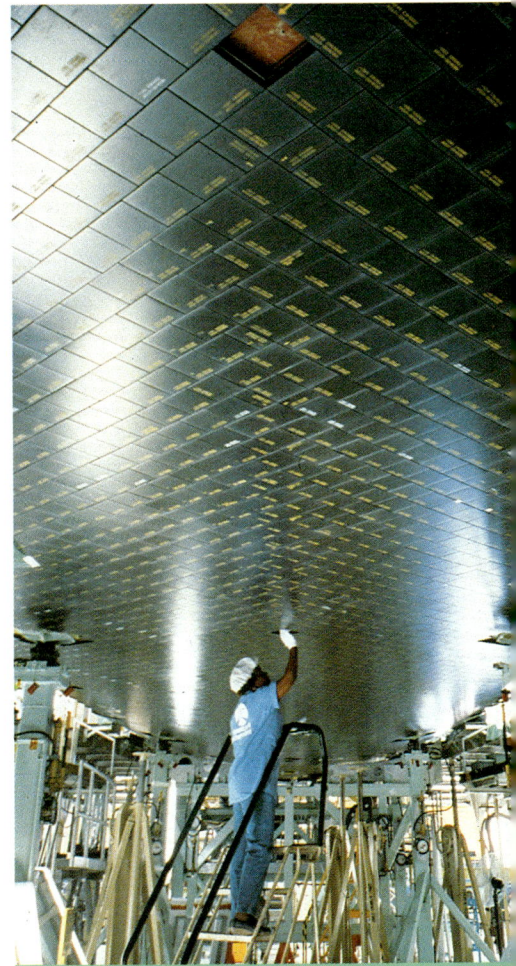

Thermal tiles protect the space shuttle from the heat of re-entry into Earth's atmosphere. What temperature do you think these tiles can withstand without melting? (Answer on page 32.)

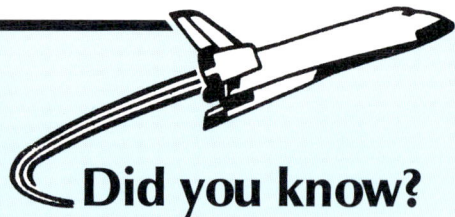

People may land on Phobos by the year 2003. Phobos is a tiny moon of Mars. The trip to Phobos is one of many plans suggested by NASA workers in the Office of Exploration. It would take about nine months for a spacecraft to travel to Phobos from Earth.

COMING ATTRACTIONS

Imagine packing your bags for a trip into space. What would you bring for a nine-month voyage? You would bring everything you needed! In space, you wouldn't be able to stop at the store for food. You wouldn't be able to send back to Earth for more air.

A space trip depends on careful planning and a lot of technology. Every possible need must be predicted. Nothing can be left to chance.

What's Next?

What long space journeys are planned for the future? Today, space planners are looking at several ideas for future trips. They call these plans **scenarios**. One scenario is a trip to Mars.

This is an artist's idea of what people working on Mars might look like.

The Mars scenario puts people on the planet by the year 2020. (How old will you be then?) Four to eight people would make up the Mars crew. Their trip would take a total of 14 months. They would spend about 20 days of that time on Mars.

Many problems must be solved before the Mars trip. Some of the technology used to go to the moon will be useful on the Mars trip. There will be some big differences, however. One is that Mars, unlike the moon, has a thin atmosphere. Huge dust storms blow across the planet's surface. The winds and blowing dust could harm equipment and cause other problems. Another difference is that the water on Mars is in the form of ice. It may be that Mars explorers can melt, clean, and use this water.

Many problems have to be solved before people head for Mars. One of the biggest has to do with the time it will take. Everything needed for 14 months of living in space will have to go on the spacecraft. There will be no room for mistakes. People's lives will depend on careful planning.

Project Pathfinder

Imagining the future is exciting. **Space futurists** are people who imagine the future of space. They think about different ways people can explore and use space.

Long-range plans are needed to make scenarios come true. **Project Pathfinder** is part of a long-range plan. It is a program in which people are working on technology for space travel. These people are using their creativity and their knowledge. They are combining *what they know* with *where they want to go.*

Project Pathfinder workers are working on the design of vehicles that can be driven on the moon or on Mars. They are studying how people can be kept safe and healthy on long space voyages.

What's next? Nobody knows. But one thing is for sure—if space travel is next, we have to start planning for it and working toward it right now.

Many tools have helped people explore the solar system. One of the Voyager spacecraft sent back images of the planet shown in the picture. What planet is it? (Answer on page 32.)

ANSWERS TO PHOTO CUE QUESTIONS

Unit Title	Question Page Number	Answer
The Final Frontier	5	The sun
Freedom Station	7	Mars
Zero Gravity	9	Underwater, in a pool
3 . . . 2 . . . 1 . . . Liftoff!	11	The runway at Kennedy Space Center in Florida
"Home Space Home"	13	Velcro
Made in Space	15	The top one. The beads are perfect spheres.
Building Blocks	17	The roof
Computer Slaves	19	Saturn's rings
Space Spinoffs	21	Toothpaste that can be safely swallowed and doesn't foam up
Medical Marvels	23	The water hyacinth
The Cost of High Living	25	A picture of Earth
"Spaceware"	27	They filter water so it's clean enough to drink.
Help Wanted: Space Worker	29	2300° Fahrenheit
Coming Attractions	31	Jupiter